Facilities, Utilities and Cleanrooms- Commissioning & Qualification

I0416757

ISBN: 9798883668523
Imprint: Independently published
© Ian Bruce

Table of Contents

1.0 Introduction to Commissioning, Qualification And Validation

1.1 OVERVIEW

The Qualification of facilities and utilities is best managed with the creation of a qualification plan. The plan can provide a framework that outlines the qualification activities, rationales, deliverables, resources and timing. However, certain qualification activities are strongly recommended and mandated by health regulators especially within pharmaceutical biotech, MedTech and medical device sectors. The regulatory legislation pertaining to the specific products and markets can inform the essential qualification requirements. Medical devices range in their principal mechanism of action, complexity and intended use. For example, the facility and supporting utilities necessary for the manufacture and packing of a surgical implant differs from an Orthopedic crutch or aid. Yet again, a medicinal product or combination device such as a pre-filled syringe with a biological formulation will require aseptic techniques to be applied during the process. This controlled environment that assures sterility is supported by qualified facilities and utilities that need to function and perform consistently. Therefore, the scope and complexity of C&Q and validations must be designed based on the products manufactured and their intended purposes. With that said, there are a number of keystone commissioning, qualification and validation activities that represent best practices that are broadly applied to meet regulations. The essentials of C&Q can be specified in company (in-house) procedures or standard operating procedures. The discrete requirements required for specific projects can then be guided with the creation of a C&Q plan.

1.1.1 HIGH LEVEL UNDERSTANDING C&Q

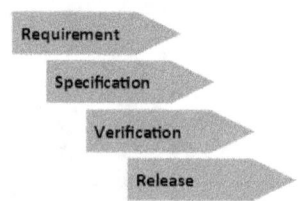

The above diagram introduces some key concepts that are applied during C&Q and validations. The intial stage deals with the requirements. Requirements should be specific, mesaable and unambigous. The requirements must decribe the use in my (intended use).

In response to requirements, a specification document is needed. The specifation document translates the user requirements into elements that help focus the contruction and design teams to meet the needs of the business. When contruction and assembly is completed the verificaiton activity can commence. Successful C&Q verification allows release of a system, factility or process for the subsequent validation activites and manufacturing.

1.1.1 REGULATORY REQUIREMENTS

If C&Q is to be applied within a pharmaceutical company, regulatory guidance in Europe is provided under Eudralex V4 Annex 15 Qualification and Validation.[1] The main stages include (i) user requirements (specification), (ii) Design Qualification (iii) Commissioning (iv) Installation Qualification, (v) Operational Qualification and (vi) Performance Qualification. The FDA and other regulatory bodies throughout the globe may require specific requirements to be considered for certain industries.

1.1.2 C&Q MODEL

C&Q Planning

Design Review

Receipt Verification

Development testing

Mechanical completion/TOP review

IV Start up

Dry loop checks

Calibration

Functional testing

IQ

OQ

---

1 Eudralex Volume 4, EU Guidelines for Good Manufacturing Practice for Medicinal Products for Human and Veterinary Use Annex 15: Qualification and Validation

The above model is an example of a common approach to C&Q. The applicable model should be created by your company as the necessary product knowledge, quality, corporate and regulatory requirements are understood best by on which elements need apply.

1.1.3 QUALIFICATION MODEL FOR MANUFACTURING SYSTEMS AND EQUIPMENT

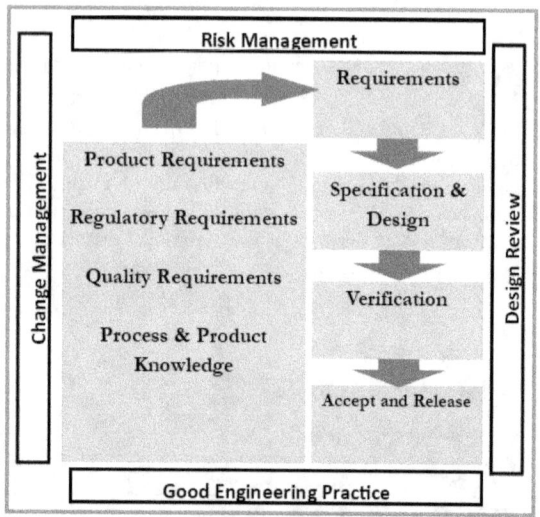

Further guidance on manufacturing systems and equipment in a pharmaceutical and biopharmaceutical context is issued by ASTM[2]

1.2 USER REQUIREMENTS SPECIFICATION (URS)

A URS is a specification document that is written for equipment, facilities, utilities or systems that defines the specific requirements to meet the use application and intended use. This includes the physical, function, operational, electrical, performance requirements and so on.

---

2 ASTM, Standard Guide for Specification, Design and Verification of Pharmaceutical and Biopharmaceutical Manufacturing systems and Equipment

Creating a written URS ensures that the user requirements are documented and approved and this can be the basis for design qualification and providing a vendor or OEMs, Original equipment manufacturers with a 'build specification' The URS is an important input during the qualification and validation process and therefore should include all critical and quality related requirements.

1.3 DESIGN QUALIFICATION (DQ)

The next element in the qualification of equipment, facilities, utilities, or systems is Design Qualification, DQ where the compliance of the design with GMP should be demonstrated and documented. The requirements of the user requirements specification should be verified during the design qualification process. The DQ should be approved by relevant roles.

1.4 COMMISSIONING

Commissioning includes the application of good engineering practices to introduce new equipment and facilities/utilizes into operation in a controlled manner that supports project delivery, safety and success.

- C&Q Planning
- Design Review
- Receipt Verification
- Development testing
- Mechanical completion/TOP review
- IV Start up
- Dry loop checks
- Calibration
- Functional testing
- IQ
- OQ

Commissioning is essentially a managed engineering approach to start up and provide turnover of equipment, utilities, systems and facilities. It involves field verification and review of system specific components and review of the construction, building and assembly of systems to ensure they meet the intended use and design specification.

After the commissioning stage, these systems are then turned over to the responsible person or owner which can then proceed with qualification and validation as required. Therefore, the real value of an effective commissioning program is reducing risks in qualification and providing the basis of success.

C&Q Planning

C&Q planning is achieved by creating a C&Q plan which sets out the activites, deliverables and key information of a project. It should cover all of the elements listed and provide guidance to the C&Q team.
It should maintained as accurate and updated if required periodically. Approval of the plan should include all stakeholders which provides a mechanism for agreeing the C&Q strategy and communicates the nuiances for each stage.

Design Review

It is best practice to include a design review of systems/facilities that are been introduced. The design review involves reviewing the proposed design (drawings, specifications, materials, configurations) against pre-appproved requirements such as the URS and other procedures (company SOPs) and standards that may need to apply.

For a direct impact system a design review is normally completed under the term Design Qualification, DQ. Design Qualification should include a review of the following documents:
- List of the approved documents in scope of the review
- Scope of DQ/review
- Attendee list and function represented
- List of open items
- List of corrective actions
- Conclusions that Design is suitable for intended use and that the project may proceed to the next stage.

Receipt Verification

Receipt Verification (RV) ensures that components or systems and assocaited ancillary items and documentation is provided and received as purchased per vendor P.O.s. RV is performed prior to installation, assembly and verification activities. Therefore is an incorrect system or equipment is provided it can be returned to the provider with minimum delay. RV checks include:

- o Confirmation of model number
- o Inspection to include equipment is not damaged
- o All specified items/components on P.O are fullfiled and reflected in delivery documentation.
  - o Documentaition and manuals are available.

Development testing

Development testing is the intial testing on a system that confirms that the system functions according to the design intent and the functions specifications. It can be completed in a simulated manner often termed 'off-line' testing. Common activities during this stage include I/O, (Input/ output)testing. Development testing, if documented accordingly following GEP and GDP standards and configuration controls may be leveraged to meet the requirements of functional testing or OQ, operational qualification testing

Mechanical completion/TOP review

Mechanical completion is confirmation that the system or equipment is phycically assembled and fixed in its use configuration. For stick built systems, drawings can be used to verify that the system is as designed and intended. Mechanical completion can also include construction activities. The contractor responsible for contruction and assembly signifies that the system is ready to handover to the C&Q team after Mechanical completion. The package of documentation is referred to as a turn-over package or TOP. Mechanical verification can occur prior to Installation verification (IV) as some checks and inspections may need access to restricted areas or areas subject to further modification. (e.g. pipr verifications prior to insulation). Open items or issues where the mechanical completion does not meet requiremetns or drawing specifications can be fixed in real time if feasible or alternatively may be tracked via a punch list. A punch list is a tracking mechanism where follow up actions, improvements or remediation is required prior to formal closure of the TOP package.

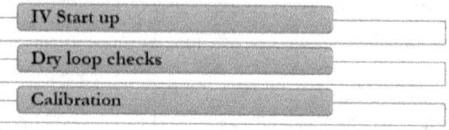

Start up is the next step after installation verification is completed for each system. The safety of the system and its operation is important to verify, so start up is done in a controlled and safe manner. A start up protocol is best practice.

The purpose of Functional testing (FT) is to verify the safety of components during operation. The activity can also be referred to as functional verification. It requires a combination of physical inspections and also completed tests in person by testing alarms, valves, interlocks, emergency stops and so on. FT of a cleanroom during the commissioning stage would include temperature and humidity monitoring, particulate monitoring and differential pressure monitoring.

IQ

OQ

Installation qualification (IQ)

For the pharmaceutical, biotech and medtech industries compliance to regulations of health authorities, competent authorities and notified bodies is required in order to manufacture and produce products. The regulators grant marketing authorizations which require companies to be subject to audit and inspection. Validation of processes, equipment, facilities and utilizes is mandated by various regulations, depending on the markets and geography. US Federal drug association require validation for both medical devices, pharmaceuticals, biotechnology, therapeutics and combination products. Installation Qualification, IQ should be performed on equipment, facilities, utilities, and systems, including computerized systems.

In respect of manufacturing facilities, cleanrooms which are built-for-purpose environments designed to control particulate and climatic conditions. The type of equipment, facility or utility should inform the requirements of IQ. However, at a minimum the following should be considered:

- o Verification of the correct installation e.g. pipe work and services as required by drawings and design specifications.
- o Verification of the correct installation according to the supplier specifications and the intended use and purpose of the system.
- o Calibration of instrumentation
- o Verification of the materials of construction.
- o Operating manual and supplier documentation
- o maintenance requirements
- o

Operational qualification (OQ)

OQ verification testing should include but is not limited to the following checks:
- o Tests that have been developed from the knowledge of processes, systems and equipment to ensure the system is operating as designed
- o Tests to confirm upper and lower operating limits, and /or "worst case" conditions

Completion of a successful OQ should allow the finalisation of standard operating and cleaning procedures, operator training and preventative maintenance requirements.

Performance qualification (PQ)

PQ should normally follow the successful completion of IQ and OQ. PQ should include, but is not limited to the following:
- o Tests, using production materials, qualified substitutes or simulated product proven to have equivalent behaviour under normal operating conditions with worst case batch sizes.

The frequency of sampling used to confirm process control should be justified; ii. Tests should cover the operating range of the intended process, unless documented evidence from the development phases confirming the operational ranges is available for manufacturing.

Re-Qualification
Equipment, facilities, utilities and systems should be evaluated at an appropriate frequency to confirm that they remain in a state of control.

re-qualification is necessary and performed at a specific time period, the period should be justified and the criteria for evaluation defined. Furthermore, the possibility of small changes over time should be assessed.

2.0 EU GMP V4 ANNEX 15

2.1 INTRODUCTION

Annex 15 provides guidelines for using a risk-based approach for the qualification and validation of pharmaceutical processes including requirements for the validation of facilities, equipment, utilities, and computerized systems.

Qualification of facilities should ensure they are suitable for their intended purpose, while the qualification of utilities should ensure they meet required specifications. The annex also discusses revalidation requirements and considers the circumstances necessitating revalidation.

Other areas of guidance includes validation documentation, control and changes and requirements for clinical trials.

2.2 CHANGES TO VALIDATED SYSTEMS OR PROCESSES

After a system or service is fully C&Q'd validation and in use for production purposes. Changes to the system should be evaluated and may require revalidation depending on the impact to the system. Replacement ancillary parts may not require validation and only maintenance intervention. However, maintenance intervention should be controlled, pre-approved and documented. A return to service checklist after any repairs of modifications is required. If upon evaluation, the change impacts the validated state of the equipment, then change management and change controls is required to preapprove the change and identify the actions needed to bring the system back into service while ensuring there is no adverse impact on the quality of the products or degradation of services provided. This detailed control and documentation is in compliance with Annex 15 which emphasizes the requirement for documentation and record-keeping throughout the validation lifecycle.

2.3 VALIDATION AND PRODUCT LIFECYCLE

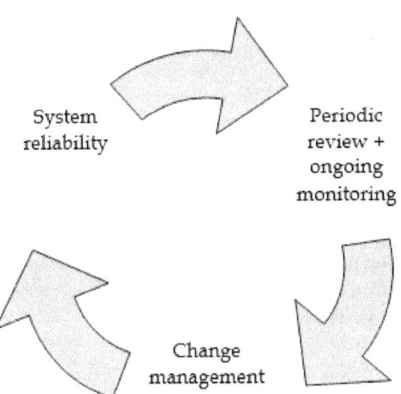

System
reliability

Periodic
review +
ongoing
monitoring

Change
management

Annex 15 emphasizes the importance of ongoing monitoring and review of validated systems. Validated systems should be periodically reviewed to ensure continued compliance and effectiveness.

3.0 CLEANROOMS

3.1 INTRODUCTION

A specially constructed enclosed area has the following controlled parameters:

- Temperature
- Humidity (Relative Humidity)
- Sound and Vibration
- Lighting
- Airflow Pattern
- Pressurization
- Particle Count
- Microbial Contamination
- Gaseous Contamination

Heating, Ventilation and Air-Conditioning (HVAC) contributes to the functioning of clean zones. It works to prevent any negative effect on production due to changes in climatic conditions. In addition, it also works to prevent product contamination and providing ergonomic working conditions.

Good Engineering Practices, application of standards, regulations and commissioning and qualification planning are necessary to deliver systems that are fit for purpose and perform as required.

- International Organization for Standardization (ISO), ISO 29463 - High-efficiency filters and filter media for removing particles in air, Parts 1 to 5.
- International Society for Pharmaceutical Engineering (ISPE) – Good Practice Guide – Heating, Ventilation and Air Conditioning (HVAC)
- International Organization for Standardization (ISO), ISO 14644 - Cleanrooms and associated controlled environments, Parts 1 to 9.

- FDA 21 CFR Parts 210 and 211 – Current Good Manufacturing Practice In Manufacturing, Processing, Packing or Holding of Drugs; General and Current Good Manufacturing Practice For Finished Pharmaceuticals
- PDA Technical Report No.13- "Fundamentals of a Microbiological Environmental Monitoring Program"
- EudraLex Volume 4, EU Guidelines for Good Manufacturing Practice for Medicinal Products for Human and Veterinary Use, Part 1, Chapter 3: Premises and Equipment)
- EN 1822 Series "High efficiency air filters (HEPA and ULPA)"

- EN 779 "Particulate air filters for general ventilation. Determination of the filtration performance."
- EN 1886 "Ventilation for buildings – Air Handling Units - Mechanical Performance"
- EN 12464-1 – "Light & Lighting of Indoor Work Places".
- ASHRAE Handbooks – Fundamentals, HVAC Systems and Equipment, HVAC Applications, Refrigeration
- ASHRAE Standard 110 – "Method of Testing Performance of Laboratory Fume Hoods"
- ASHRAE 52.2-1999 "Method of Testing General Ventilation Air-Cleaning Devices for Removal Efficiency by Particle Size"

3.2 CLEANROOM ENVIRONMENT

The environment where products are manufactured, processed and packaging can lead to contamination issue that may impact the product and safety. Therefore, an appropriate environmental cleanliness level is required to minimize the risks of particulate or microbial contamination to the product. The levels of cleanliness depends on the activity and products been provided. A cleanroom is defined as enclosed area which is environmentally controlled with respect to particles, temperature, humidity, air pressure, air pressure flow patterns, air motion, vibration, noise, viable organisms, and lighting and is designed and constructed for the intended use in mind.

ISO 14644-1[3] defines a cleanroom as *"a room in which the concentration of airborne particles is controlled, and which is designed, constructed and operated in a manner to control the introduction generation, and retention of particles inside a room".*

ISO 14644-4 A.1[4] suggests clean rooms are *"enclosed (rooms)or surrounded by further zones of lower cleanliness classification. This can allow the zones with the highest cleanliness demands to be reduced to the minimum size. Movement of material and personnel between adjacent clean zones gives rise to the risk of contamination transfer, therefore special attention should be paid to the detailed layout and management of material and personnel flow"*

Critical process areas are more stringently controlled portion of the cleanroom. Pharmaceutical cleanrooms and controlled zones should;

---

[3] ISO 14644-1 Cleanrooms and associated controlled environments Part 1: Classification of air cleanliness by particle concentration.

[4] ISO 14644-4 Cleanrooms and associated controlled environments — Part 4: Design, construction and start-up.

- Prevent the quality of products being impacted with unwanted

airborne contaminants or particles and prevent products from contaminating each other

- Provide a comfortable environment for the operators and limit

exposure to hazardous risks (e.g. drug particulates, fumes, vapors)

- Remove any contaminants form the room as effectively as

Possible and in accordance with regulatory requirements.

3.3 CLEANROOM ZONING AND CLASSIFICATION

Selecting a suitable classification for a room or manufacturing facility depends on several factors. Firstly, it can be said that sterile products require a more stringent set of criteria than non-sterile products. However, there is an extensive range of products and medical devices that are sterile but are used in different ways and consist of different materials and technology. Some sterile products are single use only and used for short-term purposes and then disposed of.

Other sterile products are used subcutaneously for longer periods or even require implantation. Therefore, the design of a facility along with its HVAC specification must be appropriate to the product being manufactured. High-risk products require greater control.

The goal of facilities and HVAC systems is to minimise contamination and the associated risks. Using a sterile versus non-sterile rule of thumb is not adequate when classifying a room or facility.

Standards including EN ISO 14644-1 and guidelines such as EU cGMP Guidelines EudraLex volume 4 Annex 1 (2008) should be consulted in order to fully understand the requirements of each ISO classification and grade of room.

ISO classifications do not specify room occupancy states but when a designation is applied, the occupancy state must be stated in the relevant documentation or procedure. The most relevant European Guideline (Annex 1 of the EU cGMP Guideline) lists four classification grades and their associated particulate limits in the 'at rest' and 'in operation' conditions. In general, for the sterile and non-sterile products, similar classes are applied, but in non-sterile production the producer could assign their classes, having similar particulate concentration, temperature, pressure etc. but lower air-change rate could be used.

The classification of a cleanroom is determined by the maximum number of particles acceptable according to a specific size and per the volume of air. The selection of the right classification for any given cleanroom needs to consider the application, the type of products been processed and the type of processes. For example, a product that can be terminally sterilized generally requires less control and can be sealed in a area that is not fully aseptic.

Particles in the air is made up of both Viable and Non-Viable Particles. Viable particles can present microbial risks. The levels of viable and non-viable particles is an indication of how 'clean' an area is. Therefore, monitoring these levels is useful in determining any adverse trends and tracking on an ongoing basis the cleanroom is operating as required.

Critical areas such as ISO Level 5 are afforded protection by areas of a lower classification. Raw materials, components and personnel are controlled with an increasing level of cleanliness in order to prevent contamination from the outside impacting the critical zones.

3.4 TYPES OF CONTAMINATION
o   cross contamination (of a product/material with another product/material)
o   non-microbial   particulate   contamination   (non-viable particles)
o   biological/microbiological   contamination   (viable particles/micro-organisms)

o   Factors Influencing Contamination Cleanliness Levels in the Manufacturing Processes:

o   process
o   air cleanliness
o   personnel hygiene and clothing
o   work practices
o   material design (material of construction, surface finishes, room finishes, equipment, open system/enclosed system, utensils etc.)
o   material cleanliness

3.5 CLEANROOM CLASSIFICATION TABLE
The maximum particle levels in per ISO 14644-1, both "At Rest" and "In Operation" particle levels are indicated below:

ISO 14644 includes the following sub-parts,

— Part 1: Classification of air cleanliness by particle concentration
— Part 2: Monitoring to provide evidence of cleanroom performance related to air
  cleanliness by particle concentration
— Part 3: Test methods
— Part 4: Design, construction and start-up
— Part 5: Operations
— Part 7: Separative devices (clean air hoods, gloveboxes, isolators and
  mini-environments)
— Part 8: Classification of air cleanliness by chemical concentration (ACC)
— Part 9: Classification of surface cleanliness by particle concentration
— Part 10: Classification of surface cleanliness by chemical concentration

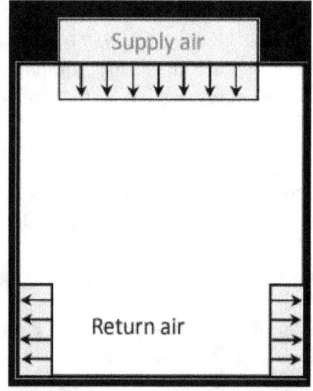

As built is the condition where the installation is complete with all services connected and functioning but with no production equipment, materials or personnel present.

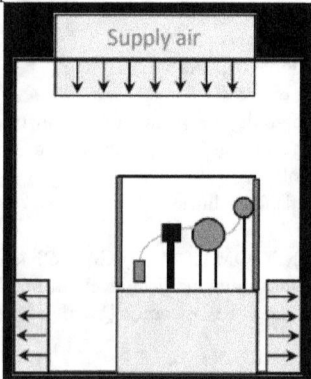

At rest condition is where the installation is complete with equipment installed and operating in a manner agreed upon by the customer and supplier, however, no personnel present.

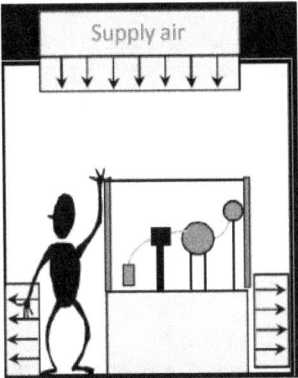

In operation is the condition where the installation is complete with equipment installed and operating in a manner agreed upon the customer and supplier and where personnel present and working.

3.6   ZONE CLASSIFICATION

Applying ISO 14644- 1 rules, cleanrooms are classified based on the level of airborne particulates within the environment. ISO Class 5-9 are summarized below, with ISO Class 9 allowing the greatest levels of particulate.

ISO ZONE 5

Critical zones are areas or clean room zones where the product, packaging, or closures are exposed to environmental conditions during the completion of the last manufacture steps. This exposure to the environment may impact product quality. This control of the critical zone is achieved by the design of the room, use of HEPA filtration during HVAV, gowning requirements, access control and the control and monitoring of conditions such as temperature, relative humidity and pressure differentials.

ISO Class 5 permits a maximum allowable particles per cubic meter: 3,520

Airborne particulates are limited to a very low level, making it suitable for environments requiring extremely high levels of cleanliness, for example, pharmaceutical production and Aseptic manufacturing and biotech.

ISO Zone 6

ISO Class 6 permits a maximum allowable particles per cubic meter of 35,200. Therefore, this cleanroom has a higher particle count compared to ISO 5 cleanrooms but are still maintained to high cleanliness standards.

ISO Zone 7

The maximum allowable particles per cubic meter is 352,000
ISO Class 7 cleanrooms have higher particle counts compared to Class 6

ISO Zone 8

ISO Class 8 permits a maximum allowable particles per cubic meter of 3,520,000. ISO Class 8 cleanrooms have even higher particle counts compared to Class 7 and are considered as controlled environments with a lower degree of cleanliness E.G. food processing, or certain medical device manufacturing processes.

ISO Zone 9

ISO Class 9 permits the maximum allowable particles per cubic meter of 35,200,000. ISO Class 9 cleanrooms are used when high levels of cleanliness is not critical.

ISO classifications ensure that cleanrooms function at the appropriate levels of cleanliness to support the operations and activities completed within them.

3.6.1 HVAC PARTICULATE CONTROL

The main purpose of the HVAC system in a cleanroom is to ensure the processing environment does not negatively impact upon product quality. Prior to the design and specification of a HVAC system, the product(s) and processes need to be understood to assess and determine the environmental controls necessary for a particular product, taking into account the type of product, the product specification and packaging and the regulatory requirements of competent authorities and notified bodies.

3.6.2 TOTAL AIRFLOW VOLUMES & RECOVERY RATES

Air change rates per hour, (AC/hr) are an important factor in contamination control. The arbitrary 20 AC/hr are a result of previous industry standards, however, nowadays, the number of air changes and depends on several factors including:

- Particle Generation Rate, (PGR) inside the space from people, equipment, etc.
- Room supply air volume
- Quality of air distribution (ventilation efficiency)
- Cleanliness of dilution air (negligible if HEPA filter are used)

3.6.3 PARTICLE GENERATION RATE (PGR)

PGR is a measure of the number of viable and non-viable particles being generated in the cleanroom from both people and equipment and to a lesser extent the building fabric as it should be designed to be non-shedding. Good cleanroom gowning and personnel training are an important factor in reducing room particle levels and AC/hr.

3.6.4 ROOM SUPPLY AIR VOLUME

The supply air volumetric flowrate to a room is not only determined by a required particle level in the room but also by several other interrelated factors;

- o Room heat gains (internal and external)
- o Number of occupants in the space and activities
- o Gowning levels
- o Moisture gain to the space from internal and external influences
- o Room leakage and differential pressure requirements

Heat and humidity gain are typically more easily controlled but should be considered by the HVAC designer.

Particle generation and removal is generally the main driver of the supply air volume and hence air change rates in cleanrooms.

HVAC designers default to "rules of thumb" for supply air rates by class of space, rather than calculating the actual airflow rate based on the activities in the room.

3.6.5 NON-UNIDIRECTIONAL FLOW & UNIDIRECTIONAL

Non-unidirectional flow uses air turbulence and dilution to mix particle contamination generated by people and machinery in the clean room. Clean Filtered air is delivered to the room through ceiling mounted air diffusers. This air mixes with the room air and removes airborne contamination through air extracts generally at low level in the walls.

For large rooms swirl diffusers induce room air vertically up to the diffuser to mix with the supply air. These diffusers create good dilution of contaminants in the room over the perforated diffuser type and may be used in rooms where there is minimum dust liberation.

However, they should be avoided in rooms where excessive dust is generated as they would add to the distribution of the dust and could be hazardous for the operators.

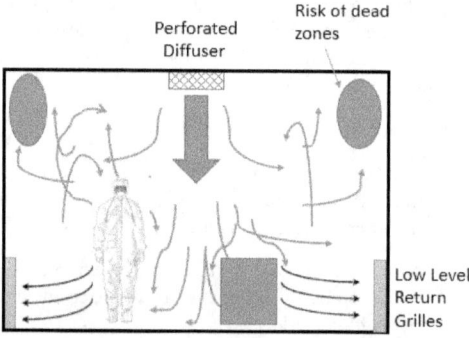

Non-Unidirectional flow with Perforated Diffuser

A shortcoming of non-unidirectional cleanrooms is the creation of air dead zones with high particle counts.

These pockets can persist for a period of time, and then disappear. This is due to currents that are set up within the room due to process related activity combined with the random nature of the downward airflow. Airflows should be planned in conjunction with operator locations, to minimize contamination of the product by the operator.

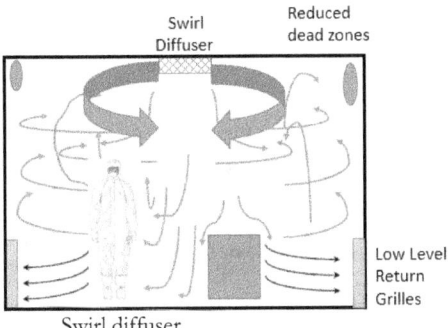

Swirl diffuser

Unidirectional flow

Unidirectional airflow (aka not laminar) is defined by ISO14644-1 as a "controlled airflow through the entire cross-section of a cleanroom or a clean zone with a constant average velocity steady velocity and air streams that are considered to be parallel".

Unidirectional airflow is achieved by supplying filtered air through 95-100% of the ceiling. The air moves vertically downward laterally from the ceiling to return air grilles at low level.

This approach allows the contamination generated by the process or surroundings to drift to the floor level where they are extracted. This is known as the displacement method as it develops minimum air turbulence.

Unidirectional airflow velocity should be uniform and sufficient to dilute and remove particles generated in the room before they settle on a surface.

The particles are finally captured by the low-level return grilles and returned through the return walls and recirculated through the filters in the AHUs or ceiling Fan Filter Units (FFU). Cleanrooms with classification rating Zone 5 or below are almost invariably designed for unidirectional airflow.

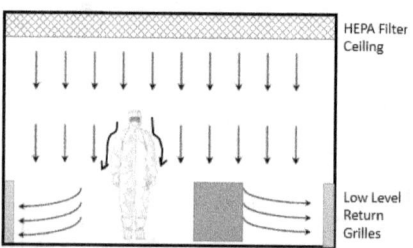

Unidirectional (turbulent or dilution) Airflow

3.6.6 AIRLOCKS

Airlocks are small rooms (typically anterooms or material transfer) with interlocked doors, constructed to maintain airflow gradient and air pressure control between adjoining rooms (generally with different air cleanliness standards).

The primary role of an airlock is to provide a pressure buffer between areas of different classification, and for that reason its own internal pressure is somewhere between (floats), or equal to, the rooms it connects.

If a door is opened, the adjoining rooms will be at the same pressure, and contamination can flow across the doorway, most easily moved by personnel or equipment dragging the contamination along.

Therefore, once the doors are closed, there must be enough air changes to reduce the contamination level before the next door opens (this is one of the few places where air changes are important (quick room recovery).

So, the higher the air change rate in the airlock, the faster the recovery and the less chance of contamination passing into the next room.

Recovery to zero counts really can't happen with personnel in the room, but the airlock presents a relatively SMALL volume of lightly contaminated air that may be dragged through the next door to be opened.

Airlocks are generally divided into two categories: Personal Airlocks (PAL) and Materials Airlocks (MAL), these control personal and material flow into and out of clean spaces through a series of interlocking doors. Airlocks also help to maintain room pressure differentials between rooms of different classifications.

There are typically three (3 No.) types of airlock pressure arrangements used:

1. Cascade - airflow from areas of higher pressure, through the airlock to the area of lower pressure.
2. Bubble – airlock is at highest pressure to surrounding rooms, air flows from the airlock to the clean rooms.
3. Sink – airlock is at lowest pressures to surrounding rooms, air flows from the clean room and corridors.

Sink or Bubble configuration are only used to restrict cross contamination of products between rooms. Cascade is the typical arrangement used where the airlock pressure floats between the pressure in both rooms.

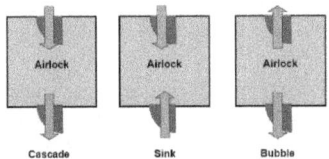

Airlock pressure configurations

Door interlocks: An interlock system or visual or audible alarm system is recommended to prevent the opening of more than one of the airlock doors at a time. Interlocked doors prevent an air flow through the airlock. Interlocks should be disarmed in case of emergency.

Pass-throughs: Small material airlock, called pass-throughs (PT) see Photo 7.2, which are too small for personnel, are used to transfer product from the higher-class rooms to lower class areas. Pass-throughs usually have interlocking sliding doors to also maintain clean room pressures between two different zones.

PT fall into two categories, namely dynamic or passive. Passive PT is tupucally used in cleanrooms with interlocked doors. PT shall be big enough to receive small items for example batch cards, samples, small consumables. Room Differential Pressure **(ΔP)**

As most facilities consist of multiple rooms with different requirements for cleanliness, differential pressure is required between the cleanrooms to ensure airflow from the cleanest spaces to the least clean spaces.

Certain operational activities may require a pressure differential to be maintained between rooms with the same classification but require an air pressure cascade (e.g. Autoclave Loading vs. Unloading). Where required this pressure must be less than the difference between room of different classification, typically +5 Pa. The requirement is to be determined at design stage and verified at certification.

The pressure differential of a cascade airlock is measured across the airlock and not across each door. Therefore, when only one door of an airlock is opened, a measurable DP between the cleanrooms persists. It also ensures the room pressure of the highest cleanliness room is maintained at a reasonable level.

Room Construction: Hard-ceiling construction is preferred for pressure-controlled spaces. In addition, air migration above the ceiling should be minimized between controlled and uncontrolled spaces.

To maintain pressurization within rooms, all doors should be fitted with continuous seals, manufactured of materials acceptable for cleanroom operation and wipe down. The gap between the finished floor and the bottom of door should be uniform at approximately 5mm (3/16 inch) when closed.

Door floor sweeps are not recommended for swing doors due to their accumulation of dirt, scratching of floor, and maintenance. Doors preferably should operate, such that the pressure differential pushes the doors closed against the frame.

Should the doors open to the low-pressure side, the door closer springs should be sufficient to hold the door closed and prevent the pressure differential from pushing the door open resulting in excessive leakage.

Air Leakage: Care shall be incorporated during design and construction to eliminate air ex-filtration within classified and controlled spaces to reduce air make-up and to maintain better static pressure control. The following, which is non-exhaustive, identifies areas where leak occurrences are most prevalent:

- Ductwork & Pipework penetrations through walls and termination into rooms
- Door perimeter
- Door closure mechanisms
- Surface Interfaces

- Spaces around equipment
- Access panels
- Electrical Fittings
- Light fixtures

The airflow leakage rate should be calculated for each room. This calculation should be based on the known architecture and the design pressure differential established for the facility

Differential Pressure Monitoring: For classified cleanrooms, it is recommended that pressure differential between cleanrooms be monitored and recorded continuously throughout each shift. There are 2 methods of measurement commonly used to monitor room pressure:

- Room to Room – Differential room Pressure
- Room to common reference point

Room to Room measurement directly meets the GMP requirement for room DP monitoring which clearly indicates the pressure relative to an adjoining room. This method is preferred for monitoring only i.e. by an EMS. When automatic room pressure control is used (VAV), the preferred measurement is room to a common reference point for stability.

There is no GMP requirement that room pressure or airflow is automatically controlled but it is recommended in clean room areas to ensure pressures are maintained. A BMS is used for automatic control and monitoring and non-critical alarming.

Fixed/manual damper control (CAV) on the supply air or return air is not recommended for cleanrooms due to its inability to compensate for fluctuations in supply and return fan output, filter loading, and exhaust modulations.

Airflow variations from dust collecting, vacuums, or process systems, and their effect on space pressurization, should be accounted for in the operation of the HVAC system.

1.1. Room Temperature and Relative Humidity

The ratio of the actual water vapour pressure of the air to the saturated water vapour pressure of the air at the same temperature expressed as a percentage.

More simply put, it is the ratio of the mass of moisture in the air, relative to the mass at 100% moisture saturation, at a given temperature.

The normal operating temperature requirement for each classification .Temperature and humidity must be appropriate to the product and process. Consideration should be made for specific product and process requirements.

A facility should meet stated relative humidity design conditions; however, the acceptability of the facility or operation depends on meeting the operating ranges.

Room Temperature and Relative Humidity (RH) requirements depend on the product requirements and operator comfort. A risk assessment should be completed to determine the criticality of temperature and humidity on product quality.

Room temperature and RH and determined both are not critical parameters as the majority of product are elastomers whose impact temperatures have a wide band. Therefore, Temperature and Humidity should only be monitored and controlled for human comfort and kept within a range to ensure human discomfort (e.g. perspiring or dehydration) does not indirectly impact product quality.

As Temperature and RH are considered non-critical parameters they should be monitored and controlled on the non-validated Building Management System (BMS). If a Risk Assessment determines that Temperature or RH are critical parameters that may impact product quality, then these parameters should be monitored by a validated Environmental Monitoring System (EMS).

It is important to define the distinction between the design and operating parameters of a space. In Figure 7.10 below, Values of Critical Parameters of a Product indicates the relationship between the design, normal operating, and qualified (validated) operating (product stability range) ranges.

3.6.7   TEMPERATURE

Room temperature is maintained for personnel comfort, considering working activities, and should take into consideration the various gowning levels worn in each area to ensure the majority of personnel are comfortable.

Energy usage should also be considered when selecting room temperature verses levels of gowning. Most HVAC systems have a lower limit of 16°C & 75%RH without additional expensive HVAC equipment with higher running costs.

The temperature sensor should be specified with an accuracy ± 1°C (± 2°F) and be fully adjustable and calibrated annually.

1.1.1.   Relative Humidity (RH)

Room Relative Humidity (RH) for personnel comfort should consider working activities to evaluate and determine the normal operating range and alert limits, while the product environmental requirements, if any, determine the qualified operating range where RH may have an impact on product. Where relative humidity levels are not specified as having product impact (e.g. aqueous product) the outer limits of the operating range or validation acceptance criteria shall be based on the following; In areas with a requirement for low particles and dust generation (e.g. clean room) the humidity should be maintained above 30% to minimize the risk of static electricity and particle generation due to dryness. Also, liquid products can lose moisture to a low-humidity space/room over an extended period. The risk of condensation and microbial growth increase above 70%RH. As a result, the Lower (Low-Low) and Upper (High-High) outer limits of the operating range or validation acceptance criteria (Alarm Limits) should be set at 30% and 70% in classified areas.

For Warehouse areas humidity levels should be maintained between 0-90%RH.

3.6.8 CONTROL AND SPREAD OF SMOKE

Systems shall not encourage the spread of smoke and fire, and in some instances, may be required to provide positive control. Careful attention must be paid to how smoke will be controlled and eliminated. It is important that smoke levels be quantified, with the necessary containment level established (i.e., cfm [m/s] smoke passage through the required smoke barrier), based on the type of structure in question, the characterization of the occupants, and their expected time to egress. Examples of positive control include pressurization of escape routes and smoke venting systems. All systems shall comply with local authorities' requirements at a minimum, and where applicable should follow the Uniform Building Code (UBC), the International Building Code (IBC), and the National Fire Protection Association (NFPA) Codes 45 and 101.

3.6.9 CLEANING

The selection of cleaning methods for cleanrooms and the sited equipment should be confirmed early in the design process because the selection may affect other design features (e.g. construction and finishes, cleanroom layouts, auxiliary services, etc.). Effectiveness of cleaning should be addressed in the validation, as applicable. Physical cleaning should be controlled by procedure and be recorded as specified.

4.0 HVAC SYSTEMS

4.1 INTRODUCTION

Heating, ventilation and air-conditioning (HVAC) provide a critical function in the manufacturing of medical devices, pharmaceutical and biotech products by contributing to the quality and environmental conditions during manufacturing, processing and packaging. Temperature, relative humidity and ventilation should be appropriate and should not adversely affect the quality of pharmaceutical products during their manufacture and storage, or the accurate functioning of equipment.

Design parameters and user requirements should, therefore, be set realistically for each project, with a view to creating a cost-effective design, yet still complying with all regulatory standards and ensuring that product quality and safety are not compromised.

4.2 HVAC SYSTEM DESIGN

The HVAC system must be appropriately selected using the specific design requirements as outlined above. The system must be able to provide clean, conditioned air to the specified areas to meet all of the quality requirements. The most important precursor to HVAC design is the comprehensive definition of the function and performance required followed by the selection of an appropriate system. A poor selection can lead to unnecessarily high-energy consumption, and operational deficiencies. HVAC systems can be divided into two main types:

All-air systems rely on the movement of large quantities of air through a central air handling unit to control room conditions, as well as provide for ventilation requirements.

They have the advantage of being relatively simple with most of the unit situated in one location; however, they are very space consuming. All-air systems tend to be relatively inflexible and not ideal for areas that are likely to need environmental alteration on a regular basis.

These HVAC systems are used for areas that have a lot of small zones, each with slightly different thermal loads but which requires constant ventilation. These systems can have poor energy efficiency if a lot of reheat is required. These are typically used in large manufacturing areas, and laboratories with many small rooms.

HVAC systems are typically situated above production areas, though this isn't always the case. Air Handling Units (AHUs) are usually located on technical floors. Air is distributed through various channels:

o   Above false ceilings

o   Through shafts (on the surface!!)

o   Through double-wall clean room walls

Design requirements shall be established in a User Requirement Specification (URS) document.

Cleanrooms may have more than one classification shared among adjoining suites, depending upon manufacturing, research and development, and containment requirements.

The cleanroom, or main assembly area, shall be specifically designated either as a specific ISO classification or Controlled Not Classified (CNC) classification, however, adjoining spaces may be designated an alternate class, and controlled via differential pressure requirements.

Other factors can affect the environmental conditions within the CR and/or CNC. Examples of these factors include the following: Number of personnel occupying each area Number and types of equipment Cleaning frequency (e.g. equipment and facility) Personnel gowning Airflow (e.g. directional, turbulent, and rate per hour) Training (e.g. movement, behavior, hygiene) Factors, such as these, may affect the cleanroom system and should be considered in the design criteria and prescribed in the user requirement specification document(s).

A well designed environment is constructed with materials that allow for ease of cleaning and sanitization. Current Good Manufacturing Practices (cGMPs) require that buildings be of suitable size, construction, and location to facilitate cleaning, maintenance, and proper operation. Additionally, the cGMPs are concerned with the potential contamination and cross contamination of product.

Based on the environmental needs of the product and/or process, controlled environment cleanrooms and areas are designed to separate manufacturing operations, and minimize the potential for contamination. The category and level of contamination control required by the product will help determine the Abbott Vascular room categorization. Abbott Vascular room classifications are based on non-viable air particulate requirements during At Rest conditions. Table 1 reflects minimum environmental specifications by Room Classification.

HEPA filters and Dehumidification

For most HVAC applications, dehumidification is best achieved by the use of cooling coils. It should be noted that dehumidification is a very high consumer of energy and should only be used if there is a real process need. When areas are not in use, the dehumidifier should be turned off, if possible.

When room humidity must be maintained below 50% during warm weather, an absorption dryer may be necessary unless the room temperature can be increased within specification to compensate.

Normal practice is to optimise size and efficiency of the absorption dryer by first removing as much moisture from the air as possible by cooling. The design of absorption dryers is normally based on a slowly rotating desiccant wheel.

Air is passed through the wheel and dried by the desiccant coating (guidance: lithium chloride especially if the wheel is not used frequently and silica gel if used permanently and with low humidity target). It is not normally necessary to size a dryer to handle the entire air volume. Drying a proportion of air and re-mixing to achieve the desired moisture content is usually sufficient.

Air humidification may be necessary during cold weather when introducing fresh air to spaces that require humidity control. When air humidification is necessary, humidifiers should be selected on the following basis:

> direct steam injection using steam
> direct steam injection using self-generative electric or gas steam humidifier.

Humidifiers should be located before the fan and the final filter which will remove any particulate generated. At least 300 mm clearance should be allowed upstream and 1 m downstream between humidifier manifolds and coils, attenuators etc. (general recommendation to be confirmed through calculation note provided by the vendor). A single manifold or multiple manifolds in parallel may be used to meet the humidification requirements as per manufacturer's recommendations.

Sound Attenuators

Sound attenuators should be provided as necessary, to achieve the specified noise levels within occupied spaces. To minimise external noise nuisance, assessment can confirm the necessity to use acoustic media (enveloped in polyester film), that is inert and corrosion-resistant at normal operating conditions. Material quality shall be equivalent to that specified for HVAC unit or ducts.
Sound attenuators should be installed in the air handling unit or ductwork. The use of sound attenuators in the air supply and air return should be based on requirements for fresh air inlet and air exhaust, and according to external noise levels that might need to be maintained at or below the ambient site noise levels.

Dampers

The provision of sufficient dampers is essential for proper control. To minimise noise transmission into the room, these should be mounted as far as possible from the diffuser.

Carefully evaluate the space-by-space pressure control that will be used in the design. Static pressure control via hard balance or dynamic control via air terminal control units are both appropriate. Consideration should be given to the overall project size, the complexity of the facility and the project budget.

Automatic volume controllers are recommended for regulating air volume independently of supply pressure. They can be selected for constant volume, variable volume or dual duct mixing applications. Automatic low-leakage fresh air and exhaust air shutoff dampers are strongly recommended to isolate the HVAC network. Fresh air dampers shall be Class 3 minimum (maximum leakage preventing coil freezing). Whenever fumigation is performed shutoff damper shall ensure Class 4 leakage rate. Where dampers are required to provide modulating control of airflow, they must be selected to provide an appropriate level of control authority. This will normally mean a damper smaller than the duct size.

Heating and Cooling

Heating mode: Low pressure hot water (LPHW) is the preferred heating medium for HVAC applications and should be used whenever practicable. Electrical heating should be avoided due to fire risk and should be limited to low power coil and in locations where no other energies are available. Hazard operability analysis (HAZOP) must be conducted if electrical heating is being considered. Cooling mode: Chilled water is the preferred cooling medium for HVAC applications and should be used whenever practicable.

The direct expansion of refrigerant in coils is an acceptable method of cooling, particularly on small isolated plants, or where lower temperatures are needed for dehumidification or for cold room. This system, however, does not normally give close control. Direct expansion coils should only be used with extreme care on variable air volume systems (if speed driver available on compressors).

Heating Coils

The face velocity of air across heating coils should not exceed 2 m/s. Coils should be made of material suitable for applicable constraints. Drains shall be located outside the casing of the HVAC unit. Coils shall be removable.

Cooling Coils

Cooling coils have been identified as potential sources of microbial contamination; therefore, careful design is required to prevent water carryover and to ensure that drain pans do not retain water. Double tube, non-welded units are recommended. The face velocity of air across cooling coils should not exceed 2 m/s. Where necessary, stainless steel or plastic eliminator blades should be provided to prevent any moisture carryover. Where provided, these must be removable for cleaning.

Ductwork

For most applications, galvanised steel ductwork will be the most appropriate form of construction; however, stainless steel or plastic construction may be necessary where there is a higher risk of corrosion due to moisture or fumes (exhaust ducts usually). Where operating pressures above 2,000 Pa are necessary, fully welded construction is recommended. For contained ducts (e.g., exhaust duct before bag-in / bag-out filter), air tightness Class C shall be followed (EN 12237). For BSL-3, fully welded construction should be considered.

Generally ductwork should be constructed to an appropriate local standard, suitable for the maximum design pressure (positive or negative), such as those published by Sheet Metal and Air Conditioning Contractors' National Association (SMACNA) in the USA, Building and Engineering Services Association (B&ES) in the UK

Where flexible connections are proposed these must be designed for the same pressure as the ductwork. Solid ducted connections are preferred for final connections to terminal HEPA filter housings. For applications where flexible connections to diffusers are used, these should be no longer than 500 mm and nominally straight.

Position	Description
1	Fresh air intake (°C, %RH, flow rate)
2	Dampers
3	Filter creating a differential pressure
4	Filter creating a differential pressure
5	Control valves for cooling fluid
6	Exhaust fan
7	Steam flow rate
8	Supply fan
9	Filter creating a differential pressure
10	Controlled room/ area
11	Extraction

Special consideration must be given to fume extract ducts where these pass through fire barriers. Using fire dampers should be avoided where the loss of extraction could make a fire situation worse. An alternative design, such as the use of fire-rated ductwork, may be necessary in these cases. A thorough risk assessment must be conducted.

Simple Representation of HVAC system

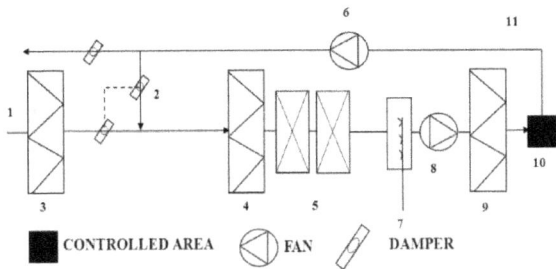

Simple HVAC diagram

4.3 ISO STANDARDS FOR CLEANROOMS

ISO-14644: Cleanrooms and Associated Controlled Environments
ISO-14644-1 Classification of Air Cleanliness
ISO-14644-2 Cleanroom Testing for Compliance
ISO-14644-3 Methods for Evaluating & Measuring Cleanrooms & Associated Controlled Environments
ISO-14644-4 Cleanroom Design & Construction
ISO-14644-5 Cleanroom Operations
ISO-14644-6 Terms, Definitions & Units
ISO-14644-7 Enhanced Clean Devices
ISO-14644-8 Molecular Contamination

For Biocontamination control the following standards apply:

ISO-14698- 1 Biocontamination: Control General Principles
ISO-14698-2 Biocontamination: Evaluation & Interpretation of Data
ISO-14698 -3 Biocontamination: Methodology for Measuring Efficiency of Cleaning Inert

4.4 TEMPERATURE

Unless otherwise required by the regulatory, product and/or process driven specifications, HVAC system design should be based on user selected temperature(s) within the range defined in Table 2. A facility should meet stated temperature design conditions; however, the acceptability of the facility or operation depends on meeting the operating ranges.

4.5 AIR HANDLING UNITS

All GMP Air-Handling Units (AHU) should have the capability of being custom designed and constructed to meet the more stringent operation and maintenance requirements for these areas. All GMP AHUs should be located in an internal plantroom to avoid risk of contamination during maintenance.

The frame shall be constructed from heavy gauge box section steel or robust aluminum penta post structure and supported on a sectional steel channel. The AHU Casing shall be constructed of modular double skin panel which shall be of cold bridge free construction. Casing panels shall be manufactured from sheet steel, with galvanized inner skins and pre-painted outer skins. Panels shall be 60mm thick and filled with CFC free water-based PU insulation foam and shall be FM approved and meet NFPA fire rating. Casing shall have a U value of less than 0.55 W/m^2K. Enclosure panels shall be manufactured from 1.3 mm galvanized sheet steel and painted PVC coated finish.

All surfaces exposed to airstream shall be hot dipped galvanized or Type 304 stainless steel as indicated on the datasheets. Aluminum will be accepted in lieu of galvanized or stainless steel

An AHU with a cooling coil shall have a drift eliminator and stainless-steel drain trays of adequate size to collect water with a minimum depth of 40mm. The drain trays shall be extended past the coils to capture all moisture carryover from the coil by the airstream. All trays shall be inclined towards the drain connection and traps adequately sized to resist the positive or negative pressure at that point in the AHU.

4.6 FILTRATION

Air filters are the primary method to reduce contamination levels in an air stream and play a very important role achieving the clean room environment. It is not only the efficiency of the filter that is important to address, but also the energy consumption (the pressure drop during the entire operation).

The air change rates should be determined by the manufacturer and designer, taking into account the various critical parameters using a risk based approach with due consideration of capital and running costs and energy usage. Primarily the air change rate should be set to a level that will achieve the required clean area condition.

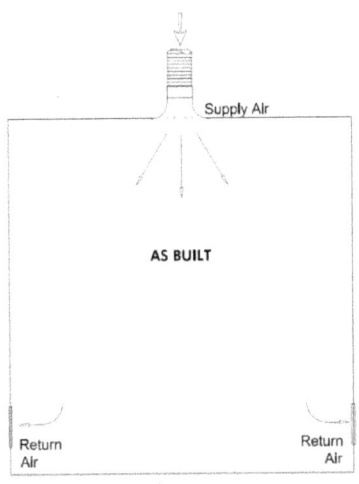

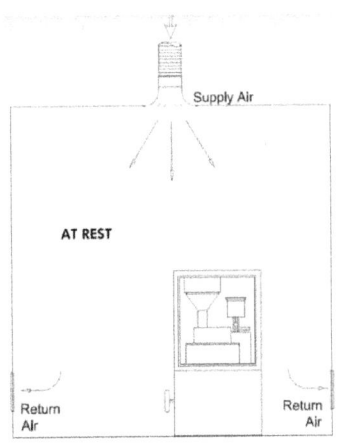

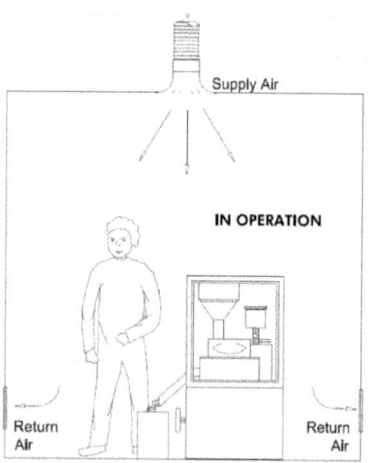

Room classification tests in the "as-built" condition should be carried out on the bare room, in the absence of any equipment or personnel. 4.1.9 Room classification tests in the "at-rest" condition should be carried out with the equipment operating where relevant, but without any operators. Because of the amounts of dust usually generated in a solid dosage facility most clean area classifications are rated for the "at-rest" condition.

Room classification tests in the "operational" condition should be carried out during the normal production process with equipment operating, and the normal number of personnel present in the room. Generally a room that is tested for an "operational" condition should be able to be cleaned up to the "at-rest" clean area classification after a short clean-up time. The clean-up time should be determined through validation and is generally of the order of 20 minutes.

Materials and products should be protected from contamination and cross contamination during all stages of manufacture for cross contamination control.

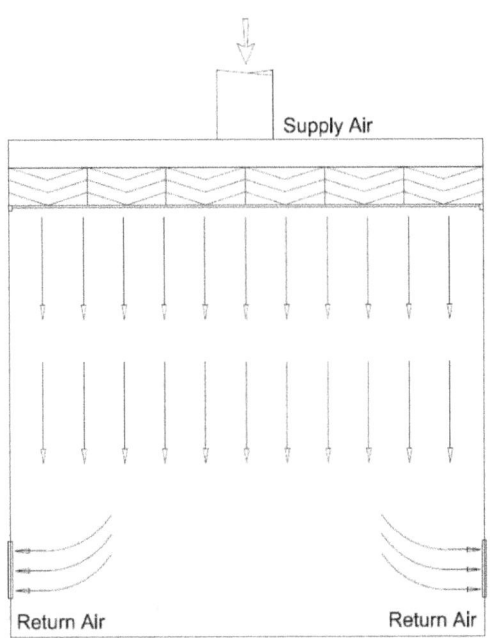

Supply Air

Return Air · Return Air

4.7 COURSE/ PRE- FILTRATION

Pre/course filtration shall be located in the AHU just after the outside and return air streams enter the recirculation unit. Level 1 filtration is the lowest efficiency, lowest cost and is used to remove all large particles (3.0 microns and larger such as insects and vegetation) found in outside air. The intention is to keep the internal components, (coils, fans etc.) and the AHU internal surfaces clean over an extended period. They also act as a pre-filtration for the Level 2 filtration and extend their life. A minimum of EN G4 (MERV 7) filters are recommended for Level 1.

Filter face air velocities shall not exceed of 2.5m/s (450 fpm). At the AHU maximum air volume flow rate, the initial pressure drop (clean) across the filter shall not exceed 100 Pa and the final pressure drops (dirty) should not exceed 250 Pa (1.0" w.g.) for panel/pleat filters as guidance, before completing the LCC analysis.

4.8 FINE / SECONDARY FILTRATION

Secondary or Fine filtration is more expensive and should be located as the last component before discharge from the AHU. This is recommended to ensure any particles or other matter (mold) generated in the AHU is captured before discharge to the ductwork and also to extend the life of filters further downstream. EN F8 or F9 (MERV 14/15/16) filters are recommended for Level 2.

Filter face air velocities shall not exceed of 2.5m/s (450 fpm). At the AHU maximum air volume flow rate, the initial pressure drop (clean) across the filter shall not exceed 100 Pa and the final pressure drops (dirty) should not exceed 450 Pa (1.4" w.g.) for panel/pleat filters as guidance, before completing the LCC analysis.

	Maximum permitted number of particles per m³ equal to or greater than the tabulated size			
	At rest		In operation	
Grade	0.5 µm	5.0µm	0.5 µm	5.0µm
A	3 520	20	3 520	20
B	3 520	29	352 000	2 900
C	352 000	2 900	3 520 000	29 000
D	3 520 000	29 000	Not defined	Not defined

maximum permitted airborne particle concentration for each grade. Showing both "at rest" and "in operation" conditions. (EU V4 Annex 1). The EU guidance given for the maximum permitted number of particles in the "at rest" column corresponds approximately to the ISO classifications.

<u>Room Air Classification (By Limits of Microbial Contamination)</u>

The HVAC systems help maintain the viable (microbial) limits within a specific area. These limits are defined in Annex 1 of the EU GMP Guide as shown below:

	Recommended limits for microbial contamination (a)			
Grade	air sample cfu/m^3	settle plates (diameter 90 mm) cfu/4 hours (b)	contact plates (diameter 55 mm) cfu/plate	glove print 5 fingers cfu/glove
A	< 1	< 1	< 1	< 1
B	10	5	5	5
C	100	50	25	-
D	200	100	50	-

5.0 COMPLIANCE TESTS FOR GMP ZONES

5.1 PARTICLE COUNT TEST

Test covers verification of cleanliness. Dust particle counts to be carried out and result printed. The number of readings and positions of tests should be defined in accordance with ISO 14644-1 Annex B5

5.2 FILTER LEAKAGE TESTS

To verify filter integrity. Filter penetration tests to be carried out by a competent person to demonstrate filter media, filter seal and filter frame integrity. Only required on HEPA filters. Refer to ISO 14644-3 Annex B6

5.3 CONTAINMENT LEAKAGE TEST

To verify absence of cross-contamination. Demonstrate that contaminant is maintained within a room by means of:
• airflow direction smoke tests
• room air pressures.
Refer to ISO 14644-3 Annex B4

5.4 AIR PRESSURE DIFFERENTIAL

This test is used to verify 'non cross-contamination'- positive air pressure pushes out particles from clean zone. Log of pressure differential readings to be produced or critical plants should be logged daily, preferably continuously. A 15 Pa pressure differential between different zones is recommended. Refer to ISO 14644-3 Annex B5

5.5 **AIR FLOW VOLUME**

To verify air change rates. Airflow readings for supply air and return air grilles to be measured and air change rates to be calculated. Refer to ISO 14644-3 Annex B13

5.6 **AIR FLOW VELOCITY**

To verify unidirectional flow or containment conditions. Air velocities for containment systems and unidirectional flow protection systems to be measured. Refer to ISO 14644-3 Annex B4

5.7 **RECOVERY**

To verify clean-up time. Test to establish time that a cleanroom takes to recover from a contaminated condition to the specified cleanroom condition. Should not take more than 15 minutes. Refer to ISO 14644-3 Annex B13

5.8 **AIR FLOW VISUALISATION**

To verify required airflow patterns. Tests to demonstrate air flows:
- from clean to dirty areas
- do not cause cross-contamination
- uniformly from unidirectional airflow units

Demonstrated by actual or video-taped smoke tests. Refer to ISO 14644-3 Annex B7

5.9 **CLEAN ROOM DESIGN CONSIDERATIONS**

5.9.1 SEASONAL VARIATIONS

All locations on earth except latitudes near the equator experience seasonal temperature changes. The changes are a consequence of Earth's orbital motion about the sun, coupled with the tilt of its axis of rotation with respect to its orbital plane. Design criteria should be based on published temperature data. The HVAC system design should consider the following:

<u>Standard Operating Conditions</u>: These are climatic conditions against which the systems must be designed to operate, control, and maintain required conditions. (These may be based on published data, which are only exceeded 2.5% or 1% of the time).

<u>Extreme Operating Conditions</u>: These are climatic conditions against which the systems must be designed to operate, without manual intervention, and without damage to the systems or the facility. Based on product / process risk assessments, extreme or standard conditions shall be used for HVAC design for dedicated areas.

<u>Location</u>
Based on the building layout, footprint and design intent, a suitable and adequate space must be identified for HVAC location. This must include provision of chilled water, heating systems, ducts and drainage. HVAC plants must be accommodated in designated HVAC plant rooms or interstitial areas.

<u>Thermal Load</u>

Thermal load can be defined as the amount of heat energy to be removed from an inner environment by equipment (HVAC) used to maintain that environment at the design temperature when worst case external temperature(s) are being experienced. The thermal load requirement should be calculated for the following:

> Max summer conditions
> Minimum winter conditions
> High rainfall
> Standard operation
> Extreme operating conditions

5.9.2   DUST, VAPOUR, OR FUME CONTROL

Highlight areas requiring dust, vapour, gas and/or fume control on the room data sheet. These areas must be controlled to remove the possibility of product contamination and to ensure the safety of the operator and environment. Areas requiring 100% fresh air or extraction to atmosphere may require greater airflow or other measures within the room to maintain environmental conditions.

In order to meet the appropriate level of cleanliness, HVAC systems require sufficient filtration to provide "clean" air to prevent contamination of the product. Pre-filters and main filters are normally suitable for most operations; however, HEPA filters are required to prevent particulate or microbial contamination for higher-classification areas

Air Change Rates

The air change rates for each room must be calculated to be sufficient for clean-up to achieve specified particulate conditions "at rest" in static conditions after a maximum of 20 minutes from completion of operations. The actual air change rate must be chosen to satisfy the most stringent requirements including GMP, GLP, heat gain, ventilation requirements and/or occupancy, including an appropriate safety factor.

The air change rate must be optimised for energy savings; however, specific attention must be paid to air locks where a greater air change rate must be applied. Air changes can be reduced (e.g. setback modes) in some circumstances ("at rest" mode, with no production activity and no personnel interventions).

Room Exhaust

Where there is a risk of active compounds being present in extracted air, filters should be fitted, preferably in the room, to prevent contamination of ductwork and the environment. The filters must be selected based on the particle size distribution of the products to be handled.

Dust Extraction and Collection

It is essential to capture dust as close as possible to the point of generation without affecting the process. In most cases dust capture should be within 100mm from the point of release. Air velocity is the key parameter in dust capture.

Pharmaceutical and chemical applications have specific collection requirements as any dust build-up in the system is likely to be of a pharmacologically active nature, sensitising, toxic and/or corrosive. It is vital to maintain transport velocities and minimise any potential for cross contamination.

A typical system should have a minimum transport velocity of 18 m/s, but this may need to be higher if heavy particles are to be collected. This velocity must be maintained throughout the system to prevent dust from dropping out in the ducts.

The dust collection must be configured with the hazardous nature of the dust in mind. A clearly defined disposal procedure for the collected dust (e.g. bag-in / bag-out system for filter and dust bin) needs to be understood at the design stage. HVAC unit shall meet EN 1886 and EN 13053 requirements.

Fans

Certified performance curves are required to verify correct fan operation. Fans that may be subjected to high temperatures, humidity, corrosive fumes or other hazardous atmospheres should be constructed using non-reactive, non-corrosive, suitable and approved materials (such as epoxy painting). Whenever H2O2 or other disinfection application is planned, material compatibility certificates shall be supplied by the vendor.

Fans must be selected to supply the design volume, taking into account the assumption that filters are half clogged, except for the terminal filter which shall be considered to be fully clogged according to EN 13053. If the terminal filter is HEPA, clogging shall be considered according to EN 1822 and the target volume is 80% of the given maximum clogged specified value.

Filtration

Face-fitting filters shall be used in all cases, as slide-in filter elements never give a good seal. The installation must be such that the airflow pushes the filter against the seal. The face velocity across the filter section shall not exceed 2 m/s. For ventilation and air conditioning applications, two minimum filtration stages are required. For certain applications, return air filtration will be required to contain highly active materials (e.g. viruses or potent compounds). Normally, these filters should be changed from the room side. However, since those filters must be integrity tested, it is recommended to place one filter in the main return duct before the exhaust fan and design return duct network, in order to ensure tightness of the duct between the room and the filter (bag-in / bag-out filter change systems should be provided for BSL-3 areas). In case of live biological agent biocontainment, decontamination up to the filter must be proven. The grade of filter and technical solution must be selected based on the product particle size distribution and occupational exposure band (OEB) level.

6.0 UTILITY GASES & WATER

6.1 INTRODUCTION

The key utilities involved for cleaning include utilities such as water, compressed gases (air, nitrogen etc.) and the heating and cooling of process equipment. Water quality can impact the effectiveness of pre-rinsing, washing, and final rinsing. Therefore, both the water temperature and quality need to be tightly controlled and monitored. Gases are typically used in order to blowdown or blowout remaining fluids or they are used as a drying step.

The term "clean utilities" in the life science industry refers to utilities that have to fulfil quality regulatory requirements. The basis of these requirements is due to the application of the utilities in the production of products or if water (e.g. water for injection) is used in the final product, or cleaning or processing where product contact may occur.

The most common utility is water, which can be supplied in different pharmaceutical grades of purity. Purified water (PW or PUW), Highly Purified Water (HPW) and Water-for-Injection (WFI) are the most common. Water quality specifications can be found in the pharmacopeias, e.g. the US Pharmacopeia. Other clean utilities can also include clean compressed air, clean gasses (e.g. nitrogen, argon and oxygen), and clean steam.

6.2 CLEAN STEAM

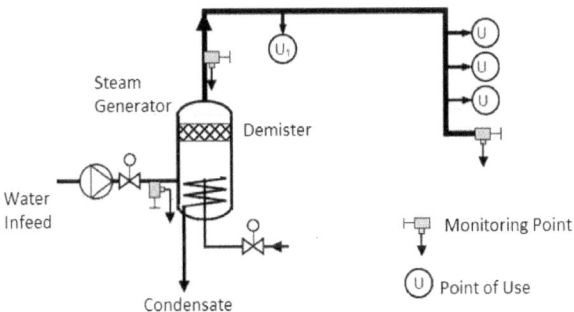

Simple Clean Steam Generation Piping and Instrumentation.

Pure steam is used in pharma and biotech for sterile application, for autoclave sterilisation etc. Distribution piping of clean steam is a critical aspect. Improper sizing of pipes may impact the production process and lead to a loss of time during sterilisation. Clean steam, also referred to as "pure steam", and gases used in manufacturing operations must be of a quality suitable for their intended purpose. The intended use of clean steam and gases must be understood in order to determine any risks to the patient or product. For example, gases that end up being part of the product must fulfil the regulatory requirements. Preventative maintenance and ongoing monitoring must be implemented for clean steam systems.

Water systems for purified water, de-ionised water and Water-for-Injection (WFI) must provide a consistent and reproducible output. Where there is moisture, there is always a risk of microbial contamination. Therefore, the design of water systems should mitigate against such risks. Good engineering practices such as using circulation loops, no dead legs and polished-surface finishes all work to provide an effective and safe system. The design should also take into account ease of sampling at the point of use. The removal of endotoxins is a requirement for WFI. Ongoing sampling monitoring the quality of water is particularly important where water systems are concerned. Procedures should be in place to ensure that effective monitoring and testing is maintained. Action limits and acceptance criteria should be clearly documented in approved SOPs or the equivalent. Failure to meet limits or acceptance criteria should initiate an investigation.

OQ Testing

Operational qualification or OQ is a formal validation activity, and as such should be completed per an approved protocol. The purpose of OQ testing is to confirm the operational and functionality of the clean steam system. This should demonstrate that all critical aspects of a URS are fulfilled. OQ verifications include:
o   Testing of temperatures and operating pressures
o   Capacity testing (under load)
o   Steam trap operation
o   Verification of automated functions and alarms
o   Check of automation systems
o   Correct function of valves and sampling points

PQ Testing

Due to the high operating temperatures and the associated lethality, clean steam systems are resistant to microbiological contamination.

Issues that arise can normally be attributed to equipment failures with the steam generator or contaminated water being supplied to the system. Bacterial endotoxin testing is used to monitor clean steam systems for both PQ purposes and throughout the life cycle of the equipment operation. Steam is condensed, sampled and tested. The condensate should meet WFI specifications with the exception of viable total aerobic count. Clean steam PQs are commonly completed using a three-phase approach to testing. The first phase ensures the system consistently operates within the required ranges and the steam provided meets the acceptance criteria. Typically, phase one bacterial endotoxin testing and physio-chemical testing is completed over a two-week period. For phase two, the same frequency and type of testing may be applied for an additional two weeks. After phase two testing, the system may be available for general use if allowed for within internal company procedures. Phase two testing at PQ should also provide a report with all results documented and reviewed. Phase three of PQ is intended to demonstrate the effective and consistent operation of the system over a longer term (approx. 12 months). Sampling is typically performed weekly.

Further Reading on Clean Steam
- o PIC/S PI009-3 – Pharmaceutical Inspection Co-Operation Scheme - Inspection of Utilities
- o EN 285 – European Standard - Sterilisation, Steam Sterilisers, Large Sterilisers
- o USP <1231> – United States Pharmacopeia - <1231> "Water for Pharmaceutical Purposes"
- o USP– United States Pharmacopeia - Monograph "Pure Steam"
- o EN 285 – European Standard - Sterilisation, Steam Sterilisers, Large Sterilisers

Schematic representation (simplified) of clean steam

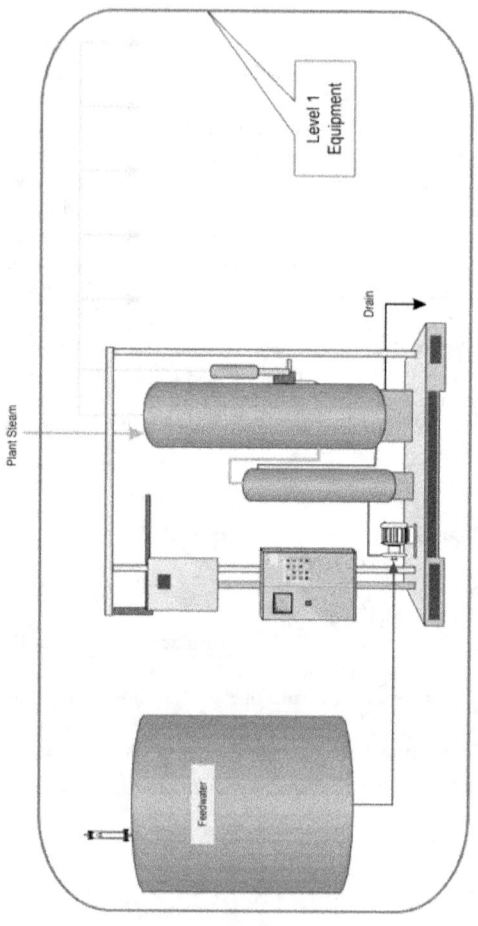

6.3 RO WATER, DI WATER AND WATER FOR INJECTION

6.3.1 WATER SYSTEMS

Water supply and the associated water systems in biotechnology and pharmaceutical plants are often critical utilities and therefore, critical to quality and safety or product. Purified water is commonly used to clean equipment and vessels, to cool or heat processing pipes and systems, in sterilize products or components via moist heat sterilsiation or indeed are used to the formulation of producing the finished product (e.g. water-for-injection). Various grades of water service a particular purpose. Some common types include:

- o RO water
- o DI Water
- o Purified water
- o Water-for injection

Reverse Osmosis, RO water and Deionized, DI water are both types of purified water. They are however, produced via different processes and therefore have different characteristics.RO water is produced by forcing water through a semi-permeable membrane. Water molecules can pass through the membrane material while larger molecules are due to their size are prevented from traversing the membrane. The RO process removes many impurities, but some dissolved solids or impurities or contaminants may still be present in the water after the RO process. DI water is created by passing water through an ion exchange membrane or material that that removes charged ions by the deionization process. Therefore, DI water seen as a more purer water that is suitable for applications in pharmaceutical manufacturing where no impurities or contaminations are desired. Critical Process Parameters for a water system include:

- o Pressure
- o pH
- o Conductivity Level
- o TOC
- o Flow rate
- o Temperature
- o Resistivity

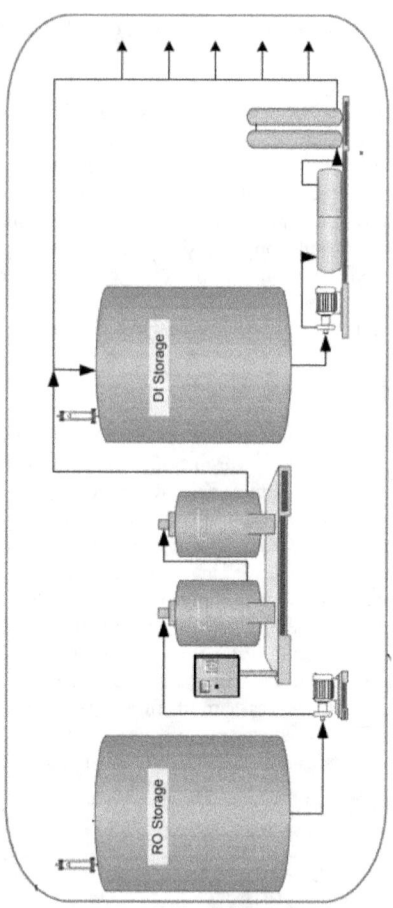

The schematic above shows the standard list of sub-systems and equipment for RO or DI water. The potential CPPs are listed above. CPPs are needed to ensure the system produces the desired quality:

The CQAs and CPPs are routinely monitored through the calibrated monitoring system which ensures any equipment failures would be detected.

- o Pressure
- o pH
- o Conductivity Level
- o TOC
- o Flow rate
- o Temperature
- o Resistivity

6.3.2   WATER FOR INJECTION:

WFI is sterile and pyrogen-free water containing no less than 10 CFU/100ml (Colony Forming Units) with a sample size of between 100 and 300 ml and an endotoxin level < 0.25 EU/ml. The use of WFI is two-fold. Firstly, it can be used for critical processing steps such as washing and rinsing. It can also be used in injectable products. WFI is a key raw material for sterile intravenous and intradermal products. WFI is produced by a Multi-Column Distillation Plant (MCDP) and must meet the microbial requirements of regulated bodies.

The cleaning of equipment, vessels and process piping is a critical activity. Any residue from a previous production batch needs to be removed in order to avoid cross-contamination. Clean in Place and Sterilize in Place skids are often utilised to allow efficient switchover between batches and/or products. Where possible, manual cleaning should be avoided unless essential due to the design of a system or particular location or configuration.

7.0 COMPRESSED AIR- GENERATION, STORAGE AND DISTRIBUTION

7.1 INTRODUCTION

Compressed air is used for valve actuation, instrument air and process air to name but a few applications. Only the point-of-use filtration and the gas quality instrumentation should be classified as level 1. When flow or pressure is a CPP, the measurement/monitoring should be performed by the system into which the gas is flowing. Additionally, the CQAs and CPPs should be routinely monitored through the calibrated monitoring system. For compressed air, the potential CPPs are listed below. For the physical system being evaluated, the use and the application of the compressed air will determine which (if not all) CPPs are needed to ensure the system produces product of the desired quality.

➢ Hydrocarbons
➢ Moisture
➢ Particulates
➢ Temperature

It is important that each point of use has appropriate sterile filters in place. If the filter is not placed directly at the point of use, control and counter measures should be implemented to address any risk of contamination downstream of the filter. Compressed air for bio-pharmaceutical use must be generated using oil free compressors with appropriate temperature controls in place.

Requirement	Clean Compressed Air (impacts product quality)	Sterile Compressed Air (impacts sterility of product)
Oil content	*Not great than 0.1mg/m³ (ISO 8573-1 Class 2)	
Microbiological requirement	Meets requirements of the environmental zone served (e.g. ISO 5)	Sterile
Filtration requirement	Minimum 0.45µm membrane filter	0.2µm membrane filter

7.2 COMPRESSED AIR DESIGN REQUIREMENTS

Compressed Air generation systems are required to address the following components in order to produce compressed air that complies to ISO 8573-1 requirements.

Class	ISO 8573-1					Viable particle counts by Air sampling Method
	Solid Particulate			Water content	Oil content	
	Maximum no. of particle per m³			Vapor pressure Dew point	Total oil mg/ m³	
	0.1-0.5 µ	0.5-1µ	1-5µ			
0	As specified by the user / supplier (≤ class 1)					100 CFU/m³
1	100	1	0	-70°C	0.01	
2	100,000	1000	10	-40°C	0.1	
3	-	10,000	500	-20°C	1	

To remind oneself of the qualification process, refer to Chapter on Qualification Requirements which specifies the qualification expectations for medicinal products. It is also an approach that can be adopted by other industries of disciplines within life sciences or Medtech.

7.2.1 DESIGN ELEMENT: INLET AIR FILTERS

Purpose: Inlet air filter is required in order to remove particles from the atmospheric air entering the compressor system.

7.2.2 DESIGN ELEMENT: AIR COMPRESSOR

Purpose: The compressor acts to compress the air into a small volume, and increases the pressure.

7.2.3 DESIGN ELEMENT: INTER COOLER
Purpose: The inter cooler lowers the temperature of hot and wet air leaving from first stage air compressor by removing water as condensate. The air then enters the second stage compression to achieve desired pressure and quality.

7.2.4 DESIGN ELEMENT: AFTER COOLER
Purpose: The after cooler lowers the temperature of hot and wet air leaving from second stage air compressor by removing water as condensate.

7.2.5 DESIGN ELEMENT :DRYER
Purpose: Dryer function is normally inbuilt in compressor and is able to eliminate any remaining moisture in the compressed air leaving from after cooler.

7.3 DESIGN REQUIREMENTS

Design elements are then translated into specific and detailed design requirements. The design requirements of compressed air generation & distribution system are specified below.

1. **General Requirements**

    i.   Capacity: of the air generation system : capacity must be calculated by determining the usage requirements of the equipment and or facilities that requires compressed air.

    ii.   Storage Vessel Capacity : to be specified

    iii.   Outlet Pressure10,00 m³/min : per equipment/ facility requirements e.g Maximum 10,00 m³/min @ 55,4 Hz Frequency, Maximum 4,81 m³/min @ 31,2 Hz Frequency

2. **Compressed Air Generation system**

    i.   Inlet Filter: meets ISO requirements

    ii.   Air Compressor Capacity : specified based on deman/requirements

    iii.   Make : Preferred make/model or similar if required

    iv.   Dryer: Inbuilt compressor unit and Heat of compression type dryer or better to produce Dew point -20°C or better as per ISO 8573

           Discharge Pressure : as recommended

           Online Dew point Sensor : with display device

           Inline Filters: 10μ, 5μ, 1μ shall be installed just after Air compressor out

3. **Compressed Air Distribution system**

    i.   Air receiver tank requirement Capacity: based on demand estimates

    ii.   Quantity: as above

    iii.   MOC: Material of construction e.g. SS304

    iv.   Drain Valve: Auto and manual type

v. Pressure Gauge:

vi. Safety Valve: As per supplier design

vii. Distribution Line   MOC: SS 304

viii. User Point /User Valve: Screw ended and manual type, MOC-SS 316

ix. Filter at user end (Process Area): $0.2\ \mu$

x. Quantity (Approx.): per requirements

xi. Compressed Air Generation System: Emergency ON/OFF switch.

xii. Safety valve on Air Receiver tank.

4. Utility Requirement

i. Available utilities Power - 3Ø (Phase), 380 – 440 VAC

ii. Documentation & Drawings

iii. Turnover packet, Design qualification (DQ), FAT, Installation qualification (IQ), Operational qualification (OQ) protocol

- General Assembly drawing of Air compressor
- Test certificate required for filter of compressor
- Test Certificate of the compressed discharge air class
- Instrument calibration certificates including copies of certificates of test equipment used in calibration.
- Critical / recommended Spare part lists with quotation

Compressed Air Distribution System Documentation

•Technical Catalogue of all Bought outs of components/instruments if installed.

•Air receiver tank and Distribution piping MOC certificate

•Leak test report of supply line and header

•Cartridge filter certificates

7.4 DESIGN QUALIFICATION

So far, in relation to compressed air generation, distribution and storage we have summarized the design elements and design requirements. (Previous sections). Both the design elements and requirements are inputs to the Design Qualification. At this point in a project an approved User requirements specification should also be available. DQ is an evaluation of the design elements and design requirements that the URS and Vendor specifications. Note: Vendor specifications are often documented in a Functional design specification (FDS) which in simple terms is an 'answer' to each of the requirements specified in the URS.

7.4.1 DQ EVALUATION

In this section the design requirements are benchmarked against the URS and Vendor responses or specifications. The Description is based on the design element and design requirement. The URS requirements are assumed to be already approved in the separate document. The vendor specification is the response of the vendor and can be a specific document that is created or alternatively if oof-the-shelf it may be a operating manual of similar document.

Description: Capacity

- o   User Requirements Specification: Generation of 1400 CFM with outlet pressure of 6-8 kg/cm2.

- o   Vendor Specification: Generation of 1600 cfm with outlet pressure of 6-8 kg/cm2.

- o Verification: While the vendor specified system has a higher capacity, this is acceptable.

Description: Inlet air filtration
- o User Requirements Specification: 3 microns with 99% efficiency
- o Vendor Specification: 3 microns with 99% efficiency

o Verification: Requirement is met by design and vendor.

Description: Compressed air generation

- o User Requirements Specification: Screw, non-lubricated oil free, air cooled.
- o Vendor Specification: screw, non-lubricated oil free, air cooled.
- o Verification: Require met by vendor

Description: Inter cooler
- o User Requirements Specification: Air or water cooled
- o Vendor Specification: air cooled type
- o Verification: Air cooled type is acceptable.

Description: After cooler

- o User Requirements Specification: Air or water cooled
- o Vendor Specification: air cooled type
- o Verification: Air cooled type is acceptable.

Description: Dryer

- o User Requirements Specification: Must be inbuilt to the compressor unit with dryer to produce dew point -20DegC or better per ISO 8573
- o Vendor Specification: Generation of 1600 cfm with outlet pressure of 6-8 kg/cm2.
- o Verification: as specified above

This process is then replicated for the remaining user requirements. A successful DQ review will ensure all design aspects are review with accepable vendor responses to the user requirements and design intent.

WATER SYSTEMS

WFI GENERATION STORAGE AND DISTRIBUTION

WFI STORAGE AND DISTRIBUTION SCHEMATIC

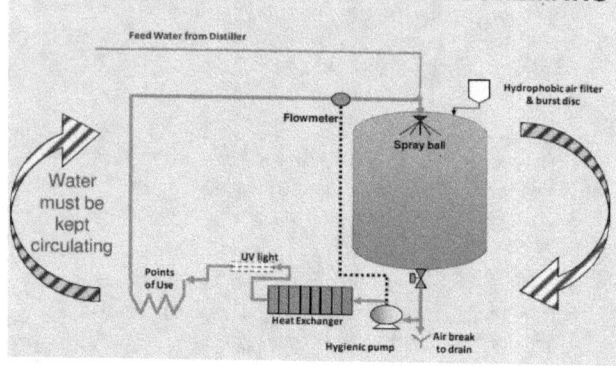

TYPICAL PW STORAGE AND DISTRIBUTION SCHEMATIC

Flowmeter controls the speed of the pump, to guarantee sufficient flow speed of water in the loop (**turbulent flow**, to prevent build-up of biofilms)

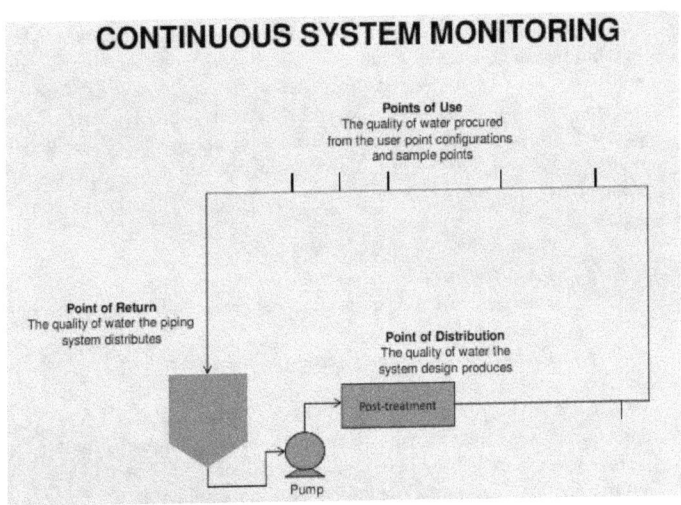

8.0 CLEAN STEAM

8.1 INTRODUCTION

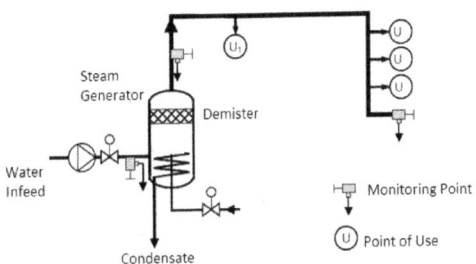

Clean Steam Generation, Piping and Instrumentation

Pure Clean Steam is used in for different functions in sterile manufacturing or used in autoclaving- moist heat sterilization. Distribution piping of clean steam is a critical aspect. Improper sizing of pipes may impact the production process and lead to loss of time during sterilisation.

Clean steam used in manufacturing operations must be of a quality suitable for their intended purpose. The intended use of clean steam and gases must be understood in order to determine any risks to the patient or product. For example, gases that end up being part of the product must fulfil the regulatory requirements. Preventative maintenance and on-going monitoring must be implemented for clean steam systems.

> ➢ Routine inspection and maintenance
> ➢ Frequency of filter change
> ➢ Frequency of the sterilisation for the gas distribution system, if applicable
> ➢ Frequency for integrity testing of the sterile filter

Water systems for purified water, de-ionised water and water-for-injection (WFI) must provide a consistent and reproducible output. Where there is moisture, there is always a risk of microbial contamination. Therefore, the design of water systems should mitigate against such risks. Good engineering practices such as using circulation loops, no dead legs and polished surface finishes all work to provide an effective and safe system. The design should also take into account ease of sampling at the point of use. The removal of endotoxins is a requirement for WFI. On-going sampling to monitor the quality of water is particularly important where water systems are concerned. Procedures should be in place to ensure effective monitoring and testing is maintained. Action limits and acceptance criteria should be clearly documented in approved SOPs or equivalent. Failure to meet limits or acceptance criteria should initiate an investigation. The potential CPPs are listed below for clean steam systems:

9.0 FACILITIES MONITORING

9.1 OVERVIEW

Control, maintenance, and system monitoring of cleanrooms must be conducted in accordance with defined standard operating procedures and includes but is not limited to the following activities:
- o BMS/FMS management, including alert and/or alarm conditions
- o Energy system management, as applicable
- o Preventative maintenance
- o Out of tolerance/event notification

o   Cleanroom stop/restart management Periodic evaluation of cleanroom environments, including, but not limited to, HEPA filtration evaluation, hood certifications, and differential pressure testing

9.2 BUILDING MANAGEMENT SYSTEMS

A Building Management System (BMS) functions as an automated control system designed to oversee various aspects of a building and its facilities, including heating, ventilation, air conditioning, security, fire protection systems, and more. This system comprises numerous Input/Output subsystems, controllers, servers, and workstations interconnected over a control network. Its primary purpose is to regulate, monitor, alert, and track equipment operations.

BMS systems are alternatively known as Facilities Management/Monitoring Systems (FMS), Energy Management Systems (EMS), Building Automation Systems (BAS), or similar terms.

Environmental Monitoring Systems (EMS) operate similarly as automated control systems. They consist of Input/Output subsystems, controllers, servers, and workstations linked over a control network. Their main function is to monitor, alert, and track critical environmental process parameters such as temperature, humidity, differential pressure, conductivity, and the status of coolers/refrigerators, among others. Here's a suggested classification of BMS and EMS systems based on their intended use.

Building Management System (BMS)	
System Classification	GxP
Data Usage	Data not used for GxP impacting decisions. Engineering use only
Monitoring	No Critical process parameters are monitored by the system
Controls	No GxP equipment
System Boundaries	Up to the point of use of the system or equipment
Validation	Not required

Environmental Monitoring System (BMS)	
System Classification	Non GxP
Data Usage	Data may be used to make quality decisions and product release decisions. Data is used to determine compliance.
Monitoring	Critical process parameters are monitored by the system
Controls	Critical alarm limits are controlled
System Boundaries	From the point of use
Validation	Not required